Refuting Evolution

the Theory of

Balance and Harmony

Evolution is a myth

Creation is Reality

By

Roy L. Waddoups

copyright Roy Lynn Waddoups

Introduction

Over the years I've found myself quite amazed that
some modern day scientists make claim that the world, and
all the life on this planet, started by pure chance. Therefore
I started researching to find out how they could make such
a claim. I found that modern science is brilliant, in fact,
in biology and many related fields, there is a tremendous
amount of very good research going on. Medicines are
being developed for many diseases with good success. A
growing understanding of DNA is leading to promising
developments in the science of genetics and helping to
overcome many maladies. All in all, I found that science on
many fronts is making some amazing advances, moving us
forward into the 21st century, and that biology research
has been very beneficial to mankind. But when I went back
to the **foundation** of the modern theory of evolution, which
proposes that this planet and all things on this planet came
to be **randomly** by pure chance, that **foundation crumbled
into the unbelievable**. Nothing that we can see with
our own eyes is happening as the theory of evolution
describes. Everything that evolutionary scientists talk

about is **just beyond our view**. The <u>foundation</u> of the theory is in stark contrast to the beautiful science that is going on all around us and, in fact, in direct opposition to the **research and achievements** of our scientists which are totally and completely accomplished through **Intelligent design, work, influence, and power.** It became clear that I should step forward and present a voice of reason on this issue. The world that we live in and all of the things that are happening here, are in sharp contrast to the theory of evolution. The theory has become accepted because of all that we **don't know** about our planet and universe, leaving open many possibilities for the human mind to contemplate. But at the same time evolution has shifted our focus away from the simple, basic things that are all around us and away from what we can experience and see with our own eyes. From a brick to a galaxy, all organized things are made the same way. How can we possibly go back from a world where all organized things are **made** by someone, to a world where things happen randomly or by pure chance?

My response is: We can't!

Table of contents

Chapter one

Who's Who in this Discussion

The two main points of view in the debate of how our planet and all the things on it came to be are:

1. Everything came to be, through a master plan, and is made by Intelligent design and influence, or

2. This planet and all things that are here, came into existence randomly by pure chance and all things evolved to their current state.

I will try to identify both groups of people on either side of the issue. As with any large or complex group of people there are always those who don't fit into either category perfectly.

Evolutionists

This group consists of many different levels of scientists **_and others_** ranging from those who are strongly committed to evolution to those who casually or passively believe in evolution.

1. The main **core** of evolutionists are a group of people, **some of whom** are scientists, who do not believe in a God or an Intelligent Creator and are looking for some other explanation of where our planet came from.

This group rejects any reference to this world being designed and organized by an Intelligent Being. They also seem to have the most power and say on what evolution is all about.

2. The next level are people who have been taught the theory of evolution their **whole lives** and have been <u>**convinced**</u> that it is sound science. This group is generally committed to evolution.

3. Then there are scientists who love the field of biology and other earth sciences and studying the miraculous things on this planet. They are scientists because they want to contribute to society and help mankind. At this time evolutionary science is the only avenue through which they can contribute.

4. Another group of evolutionists consists of scientists who are truly seeking the truth about how this planet and all life on it came to be. Some have strong doubts about evolution. Many of this group of professionals and scientists are either keeping quiet about their doubts or are systematically being booted from the institutions where they work by those in group number one!

5. The last group of evolutionists are average people who if asked whether or not they believe in evolution would answer, either passively or casually, yes.

2

These five groups do not cover every single person who

may consider evolution in one form or another.

The main group that I will refer to in this book is group number one. They are the power brokers, and the ones who have the final say regarding the science of evolution.

Creationists

There are almost as many views of Creation as there are people who believe in it. These will be categorized in a few basic groups.

1. Fundamental Christians are a somewhat diverse and loosely knit group of people who cover many different religions, and when it comes down to the details of how the Creator made this world or who the Creator is, there are slight disagreements all the way to extreme disagreements! **But there is solid agreement that there is a Creator and that He made this world and all the inhabitants of it.**

2. There is also a group of **scientists** who can see that everything on the planet has been *designed and organized* by an Intelligent power, all by natural law, and that it is **completely** in harmony with science.

3. There are other Christian religions who have stepped back to some degree from the traditional teachings of Biblical Creation and consider **some form** of evolution as possible.

4. There are many other religions that have a variety of different concepts of who the Creator is and how He made this world but who believe that this world and its inhabitants were truly Created.

5. Yet another group of people have no particular religious affiliation, but look at this beautiful and wondrous planet and can see the hand print of God on all things.

The Biblical teaching in Genesis tells us that the Creator made this planet and all the life upon it. The Bible teaches us that the Creator did something, it **does not** tell us **how** he did it! The core group of Creation Scientists have not been able to pull together and present a firm, balanced **Scientific concept** of how the Creator made the world.

The Foundation of scientific truth must be built on solid verifiable evidence.

If we **start** with an incorrect idea or concept, then, like any equation, we will **<u>never</u>** come to a **correct conclusion**.

Chapter two

The Theory of Balance and Harmony

Disharmony or disagreement seems to be the rule when it comes to all the different ideas concerning how our planet and all the things on it came into existence. Those who believe that this world was designed and organized by a Creator are not united in how it happened. Others reject the conclusions of what earth scientists have discovered over the last few hundred or so years.

On the other hand those in the scientific community also have much disagreement on how the earth came to be. Even as biological science has come under the umbrella of evolution, hundreds, even thousands, of different scientists can see that the science of evolution is built on a shaky **foundation,** and the conclusions that are being drawn, have no **verifiable** scientific **standard or basis.**

Things that are true, **don't conflict** with each other! This is why I am presenting **the Theory of Balance and Harmony! It is very simple but True**. Here is an example of what I'm talking about. You can look at just about any criminal case where there has been a robbery or murder.

As each piece of evidence comes in and is put together with other pieces of evidence, it will either fit or not fit. The evidence will or will not agree with all of the other evidence! The Truth is discovered as all of the **relevant pieces of evidence** come together, in agreement. Those things that don't fit are disregarded or removed. In other words**, if it's True, it's going to fit**! The conclusions that are made from studying virtually **every aspect** of our world will come together in harmony **if they are true!**

This book is based on the **Science of Design**, the Science of **Identifying and Gathering** materials, and the Science of **Organizing** these materials through thousands of **Processes** into the millions of things that are here, all done by Scientists, Engineers, Biologists, Physicists, Carpenters, Bakers, The Creator of the Earth, and **many more**. It is based on the fact that we have much Knowledge, but not all Knowledge, and that any future Knowledge that we find, will be in *Harmony or Agreement with the True Knowledge we already have.* **Myths, the supernatural, traditions**, and **just plain guessing** are **not** a part of the *Real Science of how this **planet** or **any thing else** was or is made.* **Verifiable Evidence** and **Pure Science** give us the **Foundation** and knowledge of how it happened! **The Theory of Balance and Harmony** keeps open, all doors and avenues of knowledge that agree with the basic and simple truths of theScience of our world that we already have. The theory of Balance and Harmony sets a solid, firm foundation of all that is certifiably true including the basic simple things that are happening around us. These **fundamental concepts** and ideas make up the "**whole**" of our planet; they are what **really Is.** With these ideas and concepts <u>Balance and Harmony</u> or agreement can be returned to our conversations about how our planet and all things on it came to be!

Chapter three

What is the Right Direction?

I **am not** suggesting in any way that Science in America should be based on the Biblical story of creation in Genesis! In fact the science that is happening across this nation and around the world is, **in many ways,** focused in the right direction. The problem is, some people in biology sciences and some of the earth sciences, have set out to **prove evolution theories** correct at any and all cost. They disregard what is **happening** on this planet.

In order to **Focus** on the most important things, I am going to <u>Underline</u>, *Italicize,* Capitalize, or **make Bold** many words that are Key to our discussion. In doing so, it is my hope that our **"Focus"** will be **brought back** to the basic Truths that are all around us. Some ideas will be presented more than once through different examples. In this book I will go over the information that is being given to us about the Origin of planet Earth and all life upon the face of it. It is about some very *basic and simple* pieces of Evidence. In fact, some are **so basic, a person** might ask why I would even bring them up! *The reason is,* many of these *most basic things* around us are being overlooked and

even set aside. Some theories go as far as disregarding virtually **everything** that **is all around us** in favor of what **may** have been. Some people are attempting to tell us that **everything we know** is backwards, and that all things, have come to be by pure chance. Even though everything around us was **made by someone**!

So in bringing forth these very **basic** pieces of evidence I am establishing what we normally call **The Simple Truth,** ideas that are well founded and which we sometimes take for granted. There is an ongoing attempt to **make relative**, **eliminate,** or **weaken** many of these very basic Simple Truths.

We must question a scientific theory that tells us that just beyond our view, all things are happening in a completely opposite way from how they happen within our view.

I will attempt to lay out what some of these forces are and to once again establish what The Simple Truth is! Of course, we don't have all knowledge about everything that has happened. We can look at *the entire Truth* about how our planet came to be as a puzzle with billions of pieces and each new piece that we find gives us a better view of how things really happened. Today we still have only a portion of the pieces of the puzzle. But when we do find new pieces of the puzzle,they will be in **harmony** or agreement with the evidence we already have and **strengthen the simple Truth** **not** weaken it.

9

Chapter four

The Foundations: Evolution and Creation

The foundation of the theory of evolution is quite simple. It is the idea, that <u>randomly, by pure chance,</u> through some great (unknown) natural phenomenon the planet was formed, and that a one celled creature was also formed and began multiplying. From this one small life form other life forms evolved consisting of billions or trillions of changes into over six million fully developed species of insect and animal life. This all started a little over four billion years ago. Plant life also started and evolved during this time. During these trillions of changes, the food supply and habitat for each of these new species that came to be, changed and evolved side by side with the insect and animal life, ***all happening randomly by pure chance* with no Intelligent design, help, direction or purpose!**

The foundation of Creation is quite simple also. It is that the universe is made up of many different elements both **seen** and **unseen.** These are the *raw materials* that this Planet was **formed** from. By **following the natural Laws** that regulate the universe and by using **many** different **processes over many years** the Creator organized this planet. He identified the **ingredients** of life and organized or formed them into all living things. He designed the physical characteristics of each *species* to coincide with the general **intelligence** of each species. (**Intelligence** is an unseen but very real **ingredient** in all living things which we will discuss later on!) Finally, through much hard work, this world, as we know it, rolled forth. Each and every step was completed through *"Pure Science"*, chemistry, geology, engineering, biology, physics, etc.

The Creator has far Greater Knowledge than we do, and with greater knowledge comes a greater ability to make things! The earth and all living things were made, very much like all the other things on our planet: **Make a plan, Design, Identify and find the ingredients**, then **Organize** them through many different **processes.** *The basics are quite simple!*

All people have a difficult time understanding Time and Space, or anything that is endless. Consequently, many people on our planet have looked at those of us who live here on this earth, as unique or an anomaly, rather than understanding that we are **a part** of the organized universe. Our planet fits, like a piece of a puzzle into the larger galaxies and universes. This world and everything that is happening here, is part of a **plan, and has a very specific reason for having been made.**

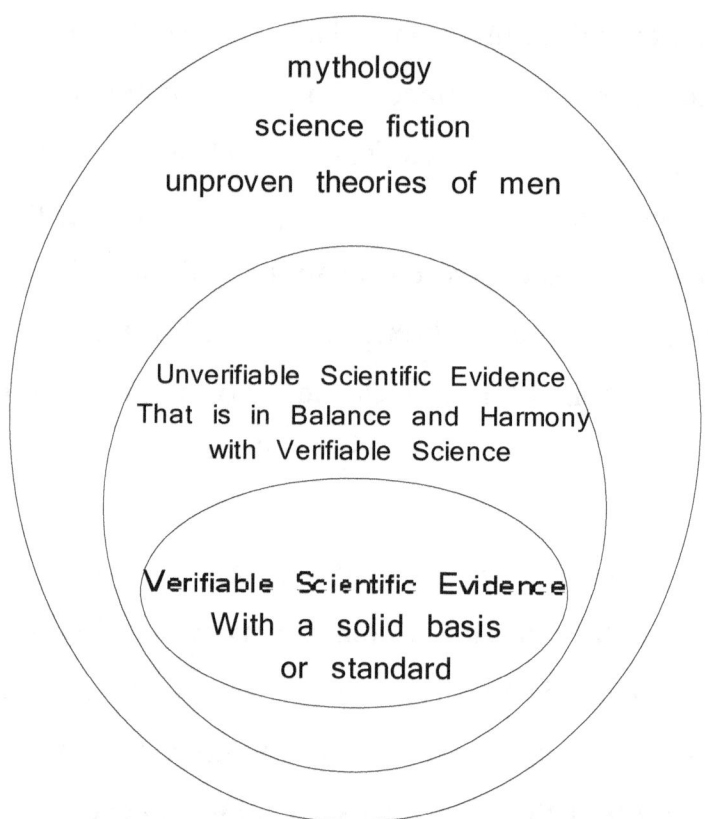

mythology

science fiction

unproven theories of men

Unverifiable Scientific Evidence
That is in Balance and Harmony
with Verifiable Science

Verifiable Scientific Evidence
With a solid basis
or standard

Verifiable Scientific Evidence with a solid basis or standard should be the **Core** of the science programs taught in our schools and universities. Ideas that are <u>in agreement</u> with this Verifiable Evidence, but not readily verifiable, should be strongly considered. Theories, guesses, science fiction, and mythology, should be considered as possible, but never taught as Factual. Today evolutionary science which is a <u>theory or guess</u>, is being taught in our schools as <u>fact!</u> Verifiable Science has been discarded! **In America we need to Return to Real Science!**

Chapter five

Perspective

Those of us who live now, are very lucky in many ways. Technology gives us the ability, not had by past generations of the human family, to see from a much larger point of view. Only three hundred years ago, a person would have been laughed at to suggest talking on some gadget to someone in Europe, South America or China. Their voices would reach only a few hundred feet at best. Today we make calls around the earth with the push of a button, even out into space. Back then, to suggest stepping on a machine and flying around the world to Australia or any other place in the world in only a few hours, would have made a person the laughing stock of his peers! But, of course, in our day and age thousands of people are doing it everyday. Three hundred years ago it may have taken months to get a letter to Australia, but now we can send a message around the earth in a few seconds with email. Millions of messages are sent each day!

We sometimes take these things for granted. *Stop now and consider!!* **We must learn from our greater perspective.** We can look back and see what others behind us could not see.

13

Will we just let this greater perspective be set aside **or, will we Learn from it**? The man who had just made a canoe thousands of years ago would likely **think it impossible** to take 101,000 tons of steel and make a boat out of it and then float it across the ocean. With our *greater Perspective* of having seen both the canoe *and* the aircraft carrier, we know that it is not only possible to float 101,000 tons of steel around the world, but that we do it all the time. Millions of people get on airplanes and fly around our planet each day. Perspective teaches us that as we look to the past and then look to where we are now, we have only seen the **tip of the iceberg** when it comes to understanding **all** that there is to know. *Much more is yet to come!*

Knowledge is the Key! Even with our modern engineering and wonderful technology, we have only just begun to learn everything there is to learn. Just because we can't see the Creator and how he does things, doesn't mean that we should doubt what can be done with **greater knowledge and power**! Like those people hundreds of years ago who would not be able to understand our airplanes or phones, we should use our perspective to know just how much **more** is possible. Why would we then, while looking at or standing on our aircraft carrier, **doubt that one with greater knowledge could Create even greater and bigger things? Greater Knowledge**

brings with it greater power to make things! We must use our better Viewpoint or Perspective to move forward in gaining this Knowledge. All of our experiences and all of the Evidence around us, teaches us that *nothing happens randomly by pure chance!* **Everything** happens when an intelligent being *does something*.

Those people who lived thousands of years ago had no idea what was possible because they <u>**could not see**</u> *what we have today. <u>We cannot see the Creator</u> or how he made this planet, but with our perspective, we* **should be able** *to understand that we don't know all there is to know and that* **much more is possible!**

Those evolutionists who place the Creator as mythology or in the supernatural, deny **billions** of verifiable facts about this world that provide a solid scientific basis of how **all** things come to be. They themselves cannot provide <u>**One**</u> **Verifiable** thing that has come to be Randomly by pure chance!

As we talk about the Creator in this book, **we are not talking** about religion or a Biblical version of the creation because the Bible only tells us that the Creator did something. *We are talking about Pure Science and* **how** *the Creator did what he did!* Everything which the creator has made, was and is made in accordance with the laws of **Pure Science** which is the **complete Knowledge** of how to make **everything** that is.

15

The more we are willing to see and use the greater perspective that we have, the **closer we will come** to finding Pure Science. The Theory of Balance and Harmony teaches us that *truth, **does not** oppose other truths*; when we have real truth it will be in agreement or harmony with all truth!

Let's take a look at **organized things** in general. We could pick any one of billions of things that are on this planet: a hat, truck, pencil, or pair of shoes. The item could be simple,such as a cake or complex like a television set. What if we said that televisions had been found in a cave and had formed themselves over a very long period of time, randomly evolving from a pile of elements and minerals into a black and white television set, and then into a high tech digital television set? What would a person's reaction be? What would be wrong with this scenario? Why wouldn't someone believe that televisions were forming themselves randomly, by pure chance and were being found in caves? Because this is not what anyone is experiencing, and it goes against what is **known to be true**! There are billions even trillions of organized things on this planet; they all, **every single one,** came to be just like a television set! <u>**Someone made them**</u>. Why would anyone even consider the concept that the over six million, unique, and complex species on this planet came to be any differently than the other organized things that are here?

Evolution

theories of men
theories of men
theories of men
theories of men
theories of men
theories of men
theories of men
theories of men
theories of men
theories of men
theories of men
theories of men
theories of men
theories of men
theories of men
theories of men
theories of men
theories of men
theories of men
theories of men
theories of men

The Foundation of Evolution is, something happened
Randomly by pure Chance With No Purpose   .

The Problem here is, there has not been
one organized item found, or evolutionary action
found, that can be Verified it happened Randomly
by pure chance with no purpose! Not One!
Its always backed up with another theory!

Chapter six

A Big Boat

Let's walk through the verifiable evidence of how ALL organized items on this planet came to be. There are many things that can help us increase our perspective and understanding. We could start with any one of millions of things that are all around us, so let's just pick one and get started. Let's go back to an aircraft carrier and watercraft. Thousands of years ago men had a problem crossing rivers and lakes. They saw logs floating and figured out that they could make a canoe by burning out a log and carving the front and back to a point. Problem solved. **The key** was gaining **Knowledge**. From that first canoe until now, thousands of different types of watercraft have been made. Let's stop here and **use our perspective** to see how all the changes were made. The steps of the process were always the same. There was a **reason, or purpose, each and every time,** that someone made something. In the process *they gained knowledge* and their *ability to create* or **make things** *was increased.* **The Creator** has a perfect **Knowledge** of how to make different things, and because of this **knowledge He has the power to**

accomplish whatever he sets out to do.

If someone had suggested back when those first boats and ships were being made, that someone could take 101,000 tons of steel and float it across the ocean, once again they would have been the laughing stock among their peers. But, of course, we know it is not only possible but that our aircraft carriers are being used around the planet everyday. Today we have **millions** of watercraft **as evidence** of where organized things come from. This historical view that we have of where all our ships came from, sheds light on how ALL things have come into existence. This is an example of evidence that is being **set aside** by those promoting evolution. Those first watercraft designers learned from their mistakes and successes, gained knowledge in the process, and then, with their increased Knowledge improved on future watercraft!

Here's the Key! **With greater Knowledge comes greater power to create,** (make a plan, design, gather materials, then organize and combine all the ingredients). Not _one_ watercraft came to be randomly by pure chance.

What is perspective? Well, you gain better perspective as you gain a better **viewpoint**. We have the luxury of seeing what has happened over thousands of years, with millions of different inventions and accomplishments. Our perspective is greater than

19

those who came before us, because our viewpoint is so much better! But evolutionists ignore the history of what men have done and dismiss virtually all of the things that we have **made** as irrelevant. They favor theories you see, that's what evolution is based upon, ideas that **can never** be verified! These theories can neither be proven right or absolutely be proven wrong. These ideas will float in relativity **forever.** Leaving science, always looking for but never finding *verifiable evidence* and so turning to **more theories** or guesses. This could go on for many years, never coming to conclusion because evolution is based on an **incorrect premise.** This premise, rather than seeing that *all of what we are doing on this planet is in perfect **harmony or agreement** with **universal truths**, denies virtually everything that we are accomplishing here! The main problem in our earth studies is, we don't see the **whole process**, but only a segment of it. The beginning of making this planet and all living things on it started many many years ago. As we rely on the simple truths around us, our understanding will grow,* **there simply are no organized things, in any place or at any time that came to be by pure chance!** *One hundred percent of organized things on this planet **that can be verified** were made by intelligent beings! There is no* <u>**scientific standard or basis**</u> *to believe that anything has ever come to be any other way!*

Chapter seven

Designed

Real Science teaches us that we observe how things happen and then draw a conclusion from what we see! We have **billions of pieces of evidence** of where organized things come from, the face of the planet is covered with them. **Why is all this evidence disregarded by evolutionists?** Why is the standard of our modern biology based on **theories rather than verifiable evidence**?

Because science has left out the Creator! As you add the Creator back into the scientific equation, everything comes back into Harmony or agreement, because **Everything** indicates **Design!** The Design of DNA, the beauty of the flowers, the wonder of a puppy or kitten, or a majestic stallion running in the wind, all of nature shows us the **hand print** of the Creator!

As Intelligent Beings, our bodies were made through the **natural processes** of creating life. We were D**esigned, with all the right tools,** to do everything we would need to do to build this world as it is! Look at your Hand, **Designed** to do whatever you need to accomplish! Look at the body of a fish, **Designed** to

21

live in water. Watch an eagle, **Designed** to fly and soar above the earth. The Creator of this earth is also a great artist. He loves beautiful things, flowers, trees, and creatures of every kind. Our world is filled with beauty. **This is not by chance**, any more than a **beautiful painting randomly appears without an artist to paint it!**

Now let's reflect for a moment and look at some of the things that we have made, following these **same patterns of design**. Our cars, **Designed** to transport us from place to place. Our telephones, **Designed** for us to communicate with each other. Our universities, **Designed** to help us gain **knowledge** and an understanding of our world. From a brick to **all** of mother nature, all things show us the hand print of "**a**" Creator.

Unfortunately, the leaders of *some* of our Institutions of Higher learning and many of our schools have **shut the doors** to any idea that is not within their **very narrow** view of evolution. Their science programs have set about *designing science to fit the views of evolutionists*! Whenever they see anything whatsoever happening, they **call** it evolution. No amount of **real** evidence or *a lack thereof has* changed their views.

The need to bring back **Balance and Harmony** into our biology studies is tremendous. Returning to a **verifiable science base** would yield **huge gains** in our research programs.

Chapter eight

How did it Happen?

Millions of people, if not virtually every person, at some point in their lives have wondered, "Where did I come from? Where did the trees, flowers, animals, and all living things come from?" Science is the tool we use to answer these questions.

Some answers are found within the many layers found in the crust of this earth. There are millions of fossil and geological records giving us some of these answers as to how our planet was **made**. In the earliest or deepest layers of the earth's crust we find more simple life forms such as amoebas and viruses. They helped to form a **foundation of mineral deposits**. Each of these layers of the earth's crust give us some **very important deposits or building blocks,** that make up all the elements **that we use each day** and that we will need into the **future**! Other groups of different life forms followed, each one bringing their **contribution** to the **forming** of the earth and adding to the **reserves of elements we would need** to make the thousands of items that we manufacture and use each day. **This is no accident!** Everything we need was **put here** *by design and had to be put* *here through many **different processes!*** This is all part of the **master plan!** There are **gaps** between the different life forms in the fossil record found in the earth's crust, because there was **no relationship between them**. Each layer seems to present new and different plant and animal life! Each of these layers of the earth is a **component or piece** of the whole **process that took many, many years to accomplish!** The key is that all of this happened through a *master plan*, following the many different **Processes** of creation.

Let's consider for a moment the building of an aircraft carrier. Would we build this super watercraft without an engine room? Or not put any electrical switches in the ship? Or neglect to put a galley in to feed the crew? It would make no sense at all to build an aircraft carrier without some of its most **essential parts**! In fact, right down to the smallest screws, paint, or light bulbs every single part is important! Do you think that an aircraft carrier would form itself **randomly**, with **every essential part** that is needed, without someone to **design and make it**? That would be ridiculous! Why would anyone believe that a planet **millions of times** more complex than an aircraft carrier would form itself **randomly** by **pure chance a**nd have all of the **essential** elements and chemical compounds that we and six million other species would need without **someone to plan,**

design, and make it? It was by design, the Creator of this planet put **every single essential element** or compound that we would need to make the millions of things that we use each day! Each mineral, chemical compound, or basic element was made through the many processes of heat, movement of the earth's crust, and by introducing millions of different plants and organisms. These living things eventually, over many, many years, helped to form all of these mineral deposits and compounds. Science **backs up** how many of these minerals were formed!

But all of this did **not** happen by chance! We build the aircraft carrier, much the same as the Creator built the planet! This planet has **every single thing** that we need, right down to the smallest detail, and that includes the right food and habitat for the over six million other species that are here with us!

What would we do without all of the minerals and chemical compounds that are here? Virtually everything we make on this planet depends on these hundreds of minerals and compounds! As far as we know none of the other planets in our solar system have **all** of these important elements! We have them because they were put here for our use. **This planet was** <u>**designed**</u> to have all of the things that we would need!

chapter nine

We Can't See Him

Today, evolutionists **claim** they have overwhelming evidence of evolution! But **only** after **eliminating** Intelligent Design and Creation, which accounts for virtually **one hundred percent** of the organized things that are all around us which are <u>verifiable</u>! When we add Intelligent Design back into the Scientific equation, virtually all evidence supports the **reality** that **everything** is **designed** and then **made** or **organized** from the raw materials of this planet or the universe through different **processes**! *We must question scientific theory that tells us that just **beyond our view** all things are happening in a completely **opposite way** from how they happen within our view.*

All things are made from the dust of the earth! No, we can't see this great Creator, but **All things proclaim design and organization!** No, we don't have all the answers about the Creator, but that is not reason enough to abandon a **firm scientific basis** that forms a rock solid **foundation** of understanding!

We can't see the Creator as we see each other, but we can see his **influence**, in a world that is designed and organized right down to food and habitat for **every single one** of over six million living species of creatures, including ourselves!

The Creator of heaven and earth Knows and **follows natural laws, rules and principles!** By his great **Knowledge** and adherence to these **rules and laws,** he then **controls and uses all the elements of the universe**! There are **no natural laws** that include *random, or pure chance organization of elements,* each *action* is **guided** by intelligent direction through a natural law, rule, or principal! This is in absolute **harmony** with **everything** that is happening on our planet. Flying, driving, manufacturing are all done by learning the natural laws that govern them and then using that knowledge to make and use all the things we have. Creating this planet and all things on it was no simple matter. Just like the building of an aircraft carrier takes time and millions of parts and processes, making this planet was a huge endeavor and when looked at correctly, the **data of science confirms** the **basic foundation** of **creation**.

The scientific community has gathered a huge amount of **understanding and knowledge** about this planet, and this library of information is **very important.** What we need to do now is **apply this knowledge correctly** and bring science back from a guessing game filled with unverifiable theories.

Men have always argued for their sometimes limited point of

view regardless of how true or untrue it is. Here is a jewel of truth: **All things that are True will always be in harmony with other things that are True!** There will always be Balance and Harmony if something is really True!

There are two (or more) polarized view points on where this earth comes from. By chance or by Creation. When we take the evidence that is *presented* from both sides of this issue, then we see that **one clear picture begins to come forth.**

All sides to this debate need to stand back and recognize that **real truth stands as a bulwark**, and should not be disregarded out of ideology, but learned and understood.

To be well-rounded in this discussion, we must talk about what the creation science position has been, also. It is quite **incomplete** when compared with all the **verifiable** scientific evidence of how all things come into existence. Those promoting creation from the viewpoint of the Bible have largely **simplified** the process and made it into a **supernatural** event. They have also **set aside** much **very relevant information** that the scientific community has learned. To gain a full understanding and knowledge of how this planet was made it may very well take **thousands of detailed volumes** of books to teach us how the earth was made and what each **process** was. **We have only a few verses of scripture telling us this story.** We all need to understand that **Truth** is in harmony or agreement with **all other Truth**. The Creator **doesn't** make natural law; He has knowledge and understanding of it, then He **Follows the Law and Uses it to accomplish His will! His Supreme Knowledge bridles, then controls, and utilizes the power of the universe.** All things are done in perfect order, using every **Law, Rule, and Principle. Thus He is God!**

The Biblical story of the Earth's Creation tells us that the Creator did something; it does not tell us **how** he did it! **The Creator does not do anything by magic, unnatural, or supernatural means**! This was not just a snap His fingers and there was a planet. We must remember that **we** only know **some** of the laws that govern the universe, but the Creator understands and **controls them all!** First of all, we must understand that for this great Creator to communicate with us would be like a professor with a PhD trying to explain to a three year old toddler about a nuclear power plant. The professor would teach the child in the most simple way about the power plant, and probably would end up describing it in story form using terms and words the child understands. The child simply does not have and cannot understand all of the knowledge about a nuclear power plant.

The information about creation given in Genesis was given

to a group of people who did not understand **How** to make a planet. This information was simply stating that the Creator had done something and described it, in a simple and basic way. His intention never was to give us a detailed understanding of the total process, only a **very** basic one.

Secondly, the **length of a day is different** on each of the billions of planets in our galaxy and the universe because of their different sizes and orbits! In the Bible the term **day** is relative to a place or situation we don't have a **full understanding of. So a "day" is a period of time.** The Creator made this planet in several major time periods through many different processes over the period of **very many** of **our** years. He used some of the same **processes of science** that we use today. Science, and Faith in the Creator, when viewed through **Pure** Knowledge and Understanding, are in perfect Balance and Harmony! **It's time for a Change**. Some of the **current** ideas of both science and religion are **incompatible** with what **is happening** on our planet and those ideas should be **scrapped**, *not defended*!

Many of the ideas that creationists have, came from the same place as the ideas of the theory of evolution, such as traditions, guesses, and theories! The understanding that Scientists have given us about how the planet was made is very **relevant and is crucial** in our Earth Science studies! But we need to stick with **solid verifiable evidence**, and not rely solely on theories or guesses.

Our Scientists play a **major role** in the many processes of **making** all the modern things on this planet. What they do is in **agreement** with how **all things are created.** Their research guides us through these many different processes as they gain knowledge and then increased ability. What are **Pure Science** and **Pure Religion**? They are the **Complete** Knowledge and Understanding of the earth and all of its inhabitants and **how** and **why** they were made. As we've mentioned, we **don't have** the Complete Knowledge of how all this happened! But we do have **some** knowledge and a lot of evidence of how all things come to be.

The Theory of Balance and Harmony teaches us that as we learn the most basic, or "Simple Truths," we can then form a rock solid **foundation** to build upon. After we have this basic, or **true foundation**, all Truth will harmonize or agree with it.

The Scientific evidence that this planet is very old is quite Solid. The elements of this planet, in fact, are very, very old and are Eternal in nature. The materials of this planet are in a constant state of change. As time moves forward, every single thing on this planet is either in a state of being organized or in the process of deterioration. Time and environment are constantly working

31

to break down all organized things. This process can be very slow or very fast, depending on what the item is made of. Paper is an example of something that comes and goes rather quickly. Stainless steel deteriorates much more slowly. Then, as these elements break down the parts are reused over and over again. Throughout this Process no materials are completely destroyed but are only *changed to a smaller size or different form.* Wood is a good example of this. When wood is burned, it simply changes from a solid to a gas, the materials staying around to be reused in making something else. Creation is simply gaining **knowledge** then **learning the processes** of organizing these basic elements.

Yes, that means everything: your hat, your cat, your car, your house, your cell phone, YOU! All made from the so called dust of the earth! One hundred percent of what is happening on our planet follows this **very same pattern;** everything is **made** from the elements of the earth! The Creator made sure all of the elements that we would need, **were here!** Our world and the things we use and make are evidences of these processes of creation. Knowledge is the Key! The more knowledge a person has, the more power he has to make things. This includes canoes, cars, aircraft carriers, t-shirts, cakes, skyscrapers, pencils, books, planets, comets, dinner, rubber bands, stars, galaxies, and puppies! All of these things are simply put together from the basic materials or elements of the universe, and each item that is made follows a pattern similar to the overall plan of how all things are made.

We are not an anomaly! Everything happening on this planet is in **perfect order** with what is happening in the universe! Look out at the billions of stars and the probable planets around them and then back to the basic simple principals of how every organized thing in our lives comes into existence, and we must conclude that all of these stars and planets were designed and <u>made</u>! The theory of **Balance and Harmony** teaches us that if we **must make** a brick, then a planet **must** also be **made**! We have billions of bricks all around us, and **all** were made by someone. On the other hand, if bricks had formed themselves randomly, then we could also expect that planets would do the same. In fact, we could choose any item on this planet and run the same experiment, take your pick! Every single thing that we can <u>**confirm**</u>, that is **formed** from the so called dust of the earth, is **formed** by a Creator! The concept that any **organized** thing has ever come into existence, randomly by pure chance, has no **verifiable** support.

Chapter ten

What You See is What You Get

Evolutionists attempt to explain away **what we as people know**, and **what we can see with our own eyes** and then tell us that this planet and all life on it *came into existence without Intelligent influence or guidance,* and that organized things just happen! They are telling us that what **we understand and experience** each day is **not** how it happened and that we should **disregard** *how everything around us comes to be* and accept *their* **unverifiable views** about the origin of life! The Theory of Balance and Harmony teaches us that **all Truth is connected.**

In other words, if a person bakes a cake right here in hometown USA, the process of baking a cake will be similar in Africa, Australia or anywhere else. At no time in the past or future will cakes form themselves randomly by pure chance. **There will always be a baker!**

The Creator, who **is** the Master Scientist, comprehends, understands, and has Knowledge of the Laws that Govern these processes of Creation!

The Question "who is this Great Creator?" has been

34

asked by almost every person born on our planet at some time or another in his or her life. The opponents of the Creator cry out, "Where is he?"and "Why, if there is a Creator doesn't He show himself?" They call Him the unknown God. **And yet** they grasp onto the *unproven theories of men.*

The **unknown big bang**! Never have we seen one! There's not **one piece** of evidence to back it up! It is a theory or **guess**, with **no** Scientific Basis **at all.**

The **unknown crystal that started all of life**! We haven't seen that one either! How about showing us just, **"One" current organized thing** on this planet that came to be *randomly by pure chance* and which can be **verified**?

When looked at objectively and with perspective, every single fossil or geological formation can be fit into one of the thousands of **processes** that it took to form this planet. Consider all the minerals and fuel that we use on this planet. Coal was formed by layers of trees and organic matter that were put here to give us this fuel! Each of thousands of different animal and plant life came, lived, and died here, fulfilling its own purpose of life and then its remains helped to form the minerals that we use each day. Each was planned and designed to provide You and me, and all other living things, with everything **we would need** to live here on this planet. **It's a process!**

35

All the different materials that we use to make anything and everything on this planet are **here** because the Creator designed them and put them into the earth's crust by the many different *processes He used in making this planet*! This concept is in **harmony or agreement** with virtually **all** enterprise on this planet.

The Questions "Who is the Creator?" and "Where is He at?" are relevant questions. The question "Where did our planet and all life on it come from?" is also a relevant question. **All** of the answers to these questions are not ours to understand, right now. There is much Knowledge we simply don't have. A more complete understanding may yet come to us, ***but*** <u>the Simple Truths and Evidences that are all around us are very sure</u>. What we see happening all around us in manufacturing, building, and even technology, shows us the processes of Creation. What we see happening all around us is the **reality** of our world and how it came to be!

Chapter eleven

Natural Law

Once again, this is how it happens! The Creator who made this planet and all Life upon it uses all of the many fields of Science! He works through *Natural Laws.* **These Laws Govern the Universe and all things in the universe. Everything that happens, happens by and through these natural laws!**

As we learned to understand and use the **Laws of flight**, we gained the ability to fly. Now, through this **Knowledge**, we can fly around the world and even into space! Because of this **Knowledge** we have the *power to accomplish **many** things!* We have learned much about communications, energy production, and using computers to store information. By **gaining knowledge** and **understanding** of the **natural laws** that govern each of these fields, we have gained the **Power** to use and benefit from them. The Creator through his greater Knowledge and understanding of Natural Laws **controls the elements** and then uses them to do whatever He needs to do. All of life and all things are made and controlled by **following** and **using** these Natural Laws. Let's look at an example.

If you hand a person a violin who has no experience playing a violin and ask him or her to play, they most likely cannot do it. If someone gives them instruction and they practice a lot, then their understanding will grow as well as their ability to play. Over time this person may become a very good violin player if he or she puts forth the effort. There are certain **rules and principles** that one must learn to master playing the violin. This basic and simple concept is true for the most **simple** things right out to the most **complex** things! **Everything** that happens, happens by following the Principles, Rules, and Laws connected to whatever it is that one is trying to accomplish. This is true for driving a car, building a jet airplane, making a cake, or just making dinner. This is also true when making a planet, or making a puppy. *Just because organizing a planet is beyond our capabilities, does not mean that planets came to be any differently than all other things!* It takes more **Knowledge** and **consequently more Power to accomplish**!

Let's remember the man who just built one of the first canoes thousands of years ago. Because he does not have the knowledge or capabilities to build an aircraft carrier, he may think that it is impossible. The question is: is it impossible to build an aircraft carrier? Not if one has the knowledge, and consequently the power that comes with it!

Is it impossible to build a planet? Not if a person has the knowledge! **Because with the knowledge comes the power.**

It is very difficult for us as people to comprehend how large the universe is or how far away something light years away is. We usually just tune it out because we don't have the ability to comprehend it. <u>We don't need to know what happened out there to understand how the universe is operated; all we need to learn is what is happening right here and now</u>.

The **basic** Foundation of knowledge is right here! How did the Creator create life? First, he **Identified** all of the **Intelligences and created a spiritual body** for every living thing. <u>Intelligence is one of the main **ingredients** that we need to consider in our science studies</u>. **All** living things show **absolute evidence** of this basic building block of life!

What is Intelligence? Intelligence is not one single thing or level. Just as there are many elements of the earth, Intelligence has many different Levels and Qualities. Each species has a different level of intelligence. Just as a person cannot reach down into the soil and find two handfuls of dirt that are exactly the same, a person cannot find any two creatures or plants that are exactly the same! Intelligence may very well have as many, or even more characteristics than do the physical elements that are all around us. **Intelligence,** as a basic **building block or**

39

ingredient is one of the main **keys** of how life is organized and made. The individual beauty and uniqueness of every single thing that the Creator has made is **very plain to see**! Look into the eyes of another person, a wife, husband, or children, anyone! *It is plain to see*! Intelligence is one of the main **Ingredients** of life. Just like **electricity,** it is one of the **Ingredients** that makes thousands of things all around the world function. Intelligence and Spirit are a much higher tech **energy source** that gives life to all living things. What is a Spirit body? A Spirit Body is simply made up of finer elements. We're not talking about ghosts or magic, we are talking about **Pure Science**! Consider the tiny particles of radiation and just how much we actually understand about them. It's possible that there are hundreds or even thousands of similar energy sources out in our universe. There are many unseen things all around us, some that we have not yet learned to identify or comprehend.

Making a plan, designing, gathering materials and then putting them together through different Processes, that is how it is done! Making a Spirit is one of the Processes of Creation. Finding, and then organizing these finer elements into a Spirit body or life energy is **basic to all of life**. We make things all the time through similar Processes! A battery, solar panel, nuclear reactor or generator, gives us **essential energy**.

This electricity is **unseen with our naked eye** but provides us with very important contributions to our world! The more complex the thing being made, the more Knowledge it takes to make it! Just as electricity can be identified, measured, and controlled by the **laws that govern it, all things can be identified, organized, and controlled by the laws that govern them.** Radiation can be identified and measured, just as many unseen gases can be identified and then controlled, and then used for different purposes. The Creator has **identified and understands** the finer **ingredients** of Intelligence and Spirit. He organized them into the living things of our world, all made through following the natural Laws that Govern our planet. The **Evidence** of Intelligence in all of Life is **Absolute**. Every living thing shows **some** level of Intelligence. It's not just a biological function; it is an **Ingredient** built in to each and every living thing by the Creator! Once again, **Knowledge** is the **Key** to creation.

Currently we may only have a few volumes of the Knowledge that it takes to create a planet, but the Scientific Knowledge that we do have gives us a foundation to build upon. Now is the time to look at this Planet and **everything that is happening here** and use all this Verifiable Evidence to get Science back to a solid Standard or Basis.

Chapter twelve

Intelligence and Element

Now we need to discuss these two very important things, Intelligence and element. They are concepts that we cannot fully comprehend. We know some of the basics and are learning more each day, but in our science studies we have not even tapped the idea of **intelligence** as a **raw material** and one of the **ingredients** used to create all of life. And we are just now learning how small elements can be broken down to.

Here is a story to help us gain perspective and an understanding of our situation. John is an electrical engineer at a regional substation for our electrical power grid. He manages the computer center that turns on or off each transmission line. He knows and controls every screen, every switch,and every gauge, of a huge room filled with equipment. One Saturday he brought his two year old son Jimmy to work with him, while he checked and made sure everything was operating smoothly. He sat him on his lap and playfully began to tell him what each screen does. Jimmy laughed and giggled throughout this game. **The question** here is, did Jimmy **comprehend** what he was being told?

Of course not. He simply does not have **enough experience** to understand what is going on in that control room! Over the last one hundred years or so, we who live on this planet have made some huge advances in technology. But this gives us **only a glimpse** of what is **actually possible**. We are much like little Jimmy, we see but don't fully understand what we see! This is **not** a reflection of our intelligence or potential, only of our **experience** and consequently of our **understanding** of what we actually **see happening**. The universe is **filled with elements** that are unorganized and also those that are organized. What we see as we look out into the sky are all the things that have been **made** (or organized) over eons of time. The universe is also **filled with Intelligence**. Intelligence is not only one thing, it is as diverse as the elements are, and has many different levels and qualities. These two groups of things are the **building blocks** or **raw materials** that everything within our site and beyond are **made of**, by the Creator or by us! The answers are all around us. Everything, is **made by someone** from the **raw materials** that are here, through many different **processes**! No, we don't have all the answers of how this planet was made or what all of these processes are. There is much we haven't learned yet, but we will never learn about these things as long as we are following the **detour** of unproven guesses of the theory of evolution. This theory is leading us on a wild goose chase. The time is truly at hand to **discard evolution** and come to a **solid foundation** for our scientific studies. As we return back to **verifiable science**, and away from incorrect ideas and conclusions, an **explosion of knowledge** will happen. You simply **cannot start** with an **incorrect concept,** like the *foundation of evolution* presents us with, and come out with a **correct conclusion**!

By gaining an understanding of the basic concept that the dust of the earth (elements) combined with intelligence, (life energy) are the **ingredients** of, and the **basis** for all of life Scientists then, will be starting from a standard that is supported by all of the living and nonliving things that are around us! We will be bringing Science back into **Harmony or agreement** with, rather than complete opposition to, all living **and** organized things on our planet!

Chapter thirteen

Where did Dinosaurs come From?

We can learn many things as we look at what happened in the past. But most of the answers we seek, we will find in what is happening right now, right here! The motions and patterns of this earth and our solar system can teach us what's happening out in the universe! Each day, the sun comes up and a new day starts. What has just happened? A Cycle! The planet made a full rotation. Each year our planet makes a full rotation or *orbit* around the sun, another Cycle! This knowledge of our solar system will lead us to understand more of how the Creator made everything and how everything is organized. If we take what we already know about our solar system and apply it to the universe, there is much we can learn. The time it takes for each planet to <u>rotate</u> in a cycle would be the length of its day. The time it takes a planet to <u>orbit</u> around its central planet or sun is its year. *Each planet's **time** is different according to its size and where it is set in the universe.*

Each spring we go out into our gardens and find the **bones or leftovers from last years crop.** They are the remains of the plants from the last orbit or cycle of our planet around our sun, a Year.

Where did all the dinosaur bones come from?

They are the leftovers from a past **Garden or cycle** as our planet moved in an orbit around the center of the universe. Each cycle is **one** of the **many** steps that it took to **complete** this planet with all of the **natural resources** that are here. Each one of these **cycles** could be thousands, or even millions of **our** years and have happened many, many, times. In the beginning, as the process of creation was started, more simple life forms were put on the earth to lay down the mineral foundation that we find here in the deeper portion of the crust of this planet. As the process of creation moved forward through time and many cycles, different and more complex creatures and plant life were put here, each living out its **time and purpose** and then helping to form more minerals and chemical compounds in the layers of the earth. The fossil records are a history of the different life forms put here by the Creator over the period of many years. Through multiple cycles each one became a part or piece of the whole process of organizing a planet. They all helped to form this world with all of the **essential minerals and raw materials** that we have today. Without these processes we would **not** have all of these essential natural resources that we have and **need.**

The Creator came back during these cycles <u>or from time to time</u>, Redesigning, Creating, and filling the planet with new life! This planet could very well have remnants or parts of other planets also, mixed in with the materials of it. None **of this happened by chance; it was all in the master plan.**

With the **Knowledge** that the creator has, he organizes the elements of the universe through many different **scientific processes**. Note: the other planets in our solar system as far as we know, **do not** have these layers of minerals and resources and do not sustain substantial life! These planets also **form** a **shield** for earth from meteorites; **this is no accident;** it is all part of the master blueprint! The solar system was designed like it is on purpose! **<u>For us, there is great Purpose in these processes</u>.** *Without an absolutely perfect plan, extremely accurate engineering, chemistry, and an understanding of all the basic sciences, this planet could not have been made! And would not exist as it is.* There is nothing magic or supernatural whatsoever in how the Creator makes all of his creations. He is, indeed, the Greatest Scientist in the Universe! As we study and learn more, our comprehension and understanding of these cycles will increase. Dinosaurs would be made the **same** as every other living thing on our planet. First, the Creator identifies the intelligence. Intelligences are identifiable just as electricity or radiation is, we simply haven't learned how yet. The Creator then gathers the ingredients that make up spirit, (smaller or finer than the physical elements), and organizes them with the intelligence into the life energy of the dinosaur. As with any other thing the Creator does there would be a specific process, or way to do this. Just as we came to understand and then make electricity **before** we made electrical appliances, this life energy of the dinosaur <u>or any other living thing</u>, was made **before** the physical body could be formed. Then, through **the basic processes of life,** He organizes the physical body and introduces the spirit or life energy into the body. One hundred percent of the evidence on our planet supports this concept. Look at how we make a car. If we didn't have an **energy source,** could we make a car? First we need **energy**, then a **plan** is made, the car is **designed**, the materials are **identified and found**, each part is **made**, then it is all **put together!** **It's quite simple,** just like the car, the dinosaur is **made** through many different **processes** from the **dust of the earth. The living things here on our planet are made the same way as everything else is made.**

Of course, there are some of the Ingredients of Life we have not been able to identify, but that is not because those ingredients don't exist; we simply don't have that Knowledge.

Chapter fourteen

We Must Reset Our Focus

Having scientific theories is an important part of the research that is being done, but we cannot let theory displace *fact, or things that are perceived to displace things that **really are**!* As we come back to a firm Scientific Basis, we will need to stop and identify a few simple facts of the theory of evolution. If there were any truth to it and everything had started from a one-celled creature, there would be billions and billions of fossils left behind to tell us the story! It would be like *reading a book,* every change would be *recorded.* **There would be no room for controversy** because there would be insurmountable evidence. There is **no insurmountable verifiable evidence**! Each life form that has been found is **unique**.

The **gaps** that have been found in the fossils of different layers of the earth are there because there is no verifiable **relation** between those layers. To develop each layer of **minerals and resources,** the Creator introduced **different** plant and animal life, each requiring a certain amount of time to fully develop. It's quite simple, just like building a skyscraper has many steps and

49

different processes, so does creating a planet! Creation Science is just **plain basic Science**! There are hundreds of **essential** minerals and chemical compounds that are on and in our planet's crust, each one is part of a **master plan** or blue print. W*e need them all*, just like each component of a skyscraper is essential to it functioning properly.

In the debate between evolution and Creation, those who promote evolution have attempted to **draw our focus away** from the **most simple Facts of our world**. Fantastic theories or guesses are used, beautiful wildlife films which have nothing to do with evolution are being used. Each one being claimed by evolution and then drawing away our **focus** from the basic and simple truths that are all around us. They discard as evidence everything on this planet as it **Really IS**. Evolution science has gone into a sort of time warp that looks for evidence from millions or billions of years ago, but ignores almost everything that is happening **now**! Now is the time to **refocus** our Science studies. **Huge** steps **forward** could be made by removing the **lead ball** of the theory of evolution from the necks of our scientists.

Chapter fifteen

Ingredients

What is the difference between a brick and a puppy? **Ingredients and** the **processes** by which they are put together. Scientists can identify the elements in a brick. Scientists can also identify the elements in the body of a puppy! Both the brick and the puppy are made up of the elements or the dust of the earth. As we look at our planet we see the physical elements, the earth beneath our feet, the trees, rocks, rivers, oceans, mountains, everything physical made up of many different combinations of ninety atoms and twenty five new atoms *created by our scientists*! These things we can touch and see. But in recent years we have found a world we can't see with our eyes.

Take for example electricity; when we look at an electrical wire we cannot see the electricity. When we look at a battery we cannot see the electricity. We can not see radiation! We cannot see many gases. There are many things that we *have* identified that we cannot see. There are also things that we have ***not yet identified*** that we cannot see, but there is evidence that they exist. One of these is the energy or life source of the puppy.

What are the differences between a brick and a puppy? As we mentioned earlier, their **ingredients** and the different **processes** by which the ingredients are put together! This is in **perfect balance and harmony** with how everything around us is made. Each item is made from the elements of the earth and put together in many different ways. **All things** suggest intelligent design! Consider the conception of, and the birth of a baby; each step is **critical** and part of a beautiful **creation process**.

The evidence that **all Living** things have some **level of intelligence** is very obvious. Intelligence, like other elements, is one of the **main ingredients** used to make the life that is on this planet!

Plants of every kind know when it is spring and time to grow. They know when it is winter and time to drop their leaves or stop drawing water from the ground. The actions of millions of insects show a level of intelligence, each performing different and important functions to keep our world in balance and harmony. Each of the animals or creatures on our planet demonstrates a unique Intelligence. Let's talk for a moment, about one of the most important creatures on our planet, **Earthworms**! Without earthworms the vegetation of our world would have a difficult time growing, our food supply would be greatly diminished. This one little creature is very valuable to both plant and animal life.

How does an earthworm know what to do? Its body is made by the Creator from the elements of the earth through the natural **processes of life,** similar to the way all living things on this planet are made. Within the earthworm's **life energy** as an **ingredient**, is a certain *level of intelligence*. This is **similar** to an engineer who is designing an electrical device, to identify the need for and then design a lower electrical voltage to fit the need. Consequently this little creature goes about preparing the soil for plants to grow, to feed **all** living things. Earthworms have great purpose for all the other creatures of the planet. Earthworms are only one of over six million different species, all of which have a **Purpose**! What are the odds that we would have earthworms to perform their **very crucial role,** in maintaining a food supply for all the rest of life on this planet without **Design and Reasoning**? It is like thinking that **crucial** parts of the engine in our cars would form randomly, by pure chance *without someone to design and make them*!

Our **Scientists are living proof of how this process of creation works.** Through their research and work, they are making many new medicines and different products to benefit the people of the world. All of this *is being done just like **all** things are made, by **Intelligent people** for a **Purpose***!

53

Chapter sixteen

Mythology and the Ability to Reason

Mythology is Created in the minds of men! **Truth** is brought to us by evidence and Facts! It has been said that the creation story in the Bible is mythology or takes us into the supernatural. Now we need to look at what is mythology and what is real. Who is a Creator? He is one who makes things, he starts with a plan, then designs the item, then gathers materials and organizes them into something. Very much like our **engineers, construction workers, miners, and scientists of many fields** whose work **is** a very big part of **making** the things we need for comfort and survival. They are the creators of everything that is all around us.

The question must be asked, is all of this real or a myth? Of course we know it is Real! All of this - our homes, cars and everything around us - creates a **standard or basis** of where and how things come to be. For example, tee-shirts. There are billions of tee-shirts on our planet and we know they were **all made by someone from the so called dust of the earth.** This is how all organized things are made.

54

Take a look around; there is **nothing** out there that is made any other way. Once again, **Truth** is a **knowledge** of what **really happened** and what **is really happening.** There is **no reason to believe** that some organized things came to be **one way,** and other things came to be in a totally **opposite way**! This concept throws reality into **disarray** and **anarchy**.

Throughout the generations of time, mythology has cropped up many times, in many different forms. Each time this has happened, it was because men **refused to believe** in the Creator of heaven and earth, **someone they could not see**. But even though we cannot see the Creator, **all things** show evidence of all the things He has made. **Purpose and organization flow** from every bee, every earthworm, and every flower. The **core** evolutionists **hate** the concept of **purpose.** It strikes against their atheist ideology, and yet everything **they themselves do,** fits firmly into a *pattern* of creation with purpose! Their books, their own laboratories, all of their studies, all of these things are done with a purpose by an intelligent being! The pattern is **timeless**, one year, one million years, one billion years. While it is true we cannot see this creator, it is absolute folly to disregard this world and everything that is happening on it, and how virtually everything around us came to be! All of the millions of different life forms on our planet have a purpose.

As each of the more than six million species of plant and animal life live out their lives an amazing thing happens. They all fit into the Eco-system, each one fulfilling its unique contribution to mother nature, and each one receiving back what it needs. Consider the size of a **project** that synchronizes millions and millions of different situations and fills the needs of each individual one. Creating this earth and all the things on it was a **massive project!** How did the Creator do it?

The same way that **all things** are made. We need to open our eyes and see the simple things around us, they form the **basis** of how <u>all</u> things were made from the dust of the earth or universe.

Everything was made for a reason. Each person and each thing on the planet **has purpose;** we are here **for a reason**! What is my Basis? What is my Standard?

Look out at everything upon the face of this planet and just try to find *one thing* <u>that was made,</u> *for no reason or purpose*! We are included; we have purpose! **This is the Reality of our world**! Mythology will always be a guess or an idea **not** based on what we can see and experience, or that does not have a sound scientific foundation. Mythology will lead us away from the sure reality of **what really is.** Mythology comes from the **minds** of men, and is blended with science fiction and it is not founded on anything that is real, or that can be confirmed.

The odds that an aircraft carrier could form itself randomly by pure chance, are billions and billions to one, or just plain impossible. The odds that a house would just randomly form itself are not much better. The odds that even a brick would randomly come to be are billions and billions to one or none. We have billions of bricks on our planet and **ALL** were made by someone. Why would we allow ourselves to be **convinced** that even more complex things such as trees, flowers or puppies occurred randomly with no one to *think about making them, then making a plan and designing them,* and finally, organizing them through multiple processes into all of the wonderful and beautiful creatures and plants that we have here to adorn this planet? The concept that any organized thing at all happens randomly by pure chance **is just plain mythology!**

It seems we have **lost** the **ability** to **Reason**! It is **urgent** that we **regain** the **ability** to **Reason.**

Chapter seventeen

To Our Scientists

For those who are **real scientists**, consider that the **science** research and the *ideology* of evolution **are in fact very different**. *Most scientists* are truly seeking positive answers to the problems and needs of society. While at the same time, **others** are promoting the **atheist ideology** of evolution. *These* ideas undercut the *foundations of freedom* and *morality in this nation* and the world! The leading away of some of our potentially best scientific minds into a *state of relativity* is happening all around us! Even as some very good science is coming out, such as genetics and the study of DNA, the rejection of a Creator works negatively in our society. Those who are scientists and millions of students, have been taught since grade school that *monkeys are mankind's ancestors* and that all of our existence is a random occurrence. **The time is now to correct our course.** Everything that is **Really True** is in **Agreement** or **Harmony** with **All** other things that are **Really True**. **This concept may make a lot of people angry, many special interests groups don't care what the Truth is because it only gets in their way.**

58

Others only care what will make them **money or bring them power.** Unfortunately, **that's why** we are in this situation to begin with. When we leave a solid Foundation and rely on theories or guesses, it's like floating in the ocean and being pushed around by every wave that comes along. Our young people are effected negatively when they are told life is one way (**a random occurrence - no purpose**) while the world they live in, is the **exact opposite.** Their individual lives are *filled with purpose*! The undercurrent of evolution *ideology* dissolves the moral basis (an understanding of right and wrong and our response to it) of our nation. It puts our students' lives into a sort of **floating relativity** where **true things** can simply be **explained away** and **replaced** with unproven theories.

Everything has Purpose! We can't always see the purpose of things, but that's only because of our **very limited** point of view! We are **not** just another link in a huge meaningless random occurrence!

This Earth was Created as a **University** for our Learning and Experience! All of our troubles, challenges, learning, victories, and existence, are **filled with purpose.** It is very important for those who are **real scientists** to understand.

First, those who love *science* and learning and can see that there is **Purpose** in it, **as in all of** life. These scientists have had their **focus,** in relation to evolution **drawn away from Real Science** and have to one degree or another lost view of the tremendous **purpose** of life, **which evolution denies!**

Second, those who are promoting the atheist ideology of evolution have the perspective that we are nothing more than a temporary link in an evolutionary wheel. These people **hate** the concept of **purpose** connected to science and life. It disrupts the proselyting of their ideology, and the **endless theories** (never having to produce anything verifiable) it brings with it.

Science, and life are full of Purpose! As much as some want to separate science and purpose in life, it simply cannot be done. It's like thinking that intelligence can be cut out of someone and still have a complete person! Or that looking for a cure for cancer is just a random thing.

Those who work in biology and related fields need to go back and take a look at what evolution *really* stands for. Evolutionary *footings* and *foundation s*tate that **we** have no long term purpose at all. And, that as we live here and then die there was **no real reason** for you and I to have been here or for our *families and friends* to have been here!

The **ideology** of evolution destroys every positive **Cornerstone** of society, the **reason to be good over bad,** the reason to seek growth over pleasure, or to sacrifice and help others, rather than to just get all we can get! The positive influence of the Creator is literally the rudder that guides society! We need to **connect all of the dots** and open our eyes, to see the *negative effects* of the **ideology** of evolution on the foundation of our society. It teaches us that the value of human life is relative to its situation. Since we are just another link in the chain of life, those who are less desirable or inconsequential can be discarded or done away with, as **with abortion**. Adolf Hitler grasped this evolutionary concept that we have **no design or purpose,** and viewed humanity as only a link of evolution to be improved upon, where individuals not of the pure race, were nothing but **weeds** to be done away with. The underlying consequences of **evolution ideology** undermine the **foundation of society** and the value of **each person and each thing** made by the Creator.

Every single nation in recent history that has turned to **tyranny**, and destruction (the Soviet Union, China, North Korea, Cambodia, and others) did so, *destroying a belief in God first*. Then they proceeded to murder over one hundred million people in their own nations.

Evolution **subtly** attempts to erase the Creator in our own nation. To what end?

Our unalienable rights can *only* be given by God!

Chapter eighteen

Inconceivable!

It is inconceivable that any person could be so totally *set off course,* and lead to believe in a world that is virtually and completely ***out of reality***! Each person rises up every morning, opens their eyes and can see their home filled with things of every kind, puts on clothes, eats prepared food, gets in a car and drives to work where each of us assist with our small part in helping to create or make all the things that are around us. Has **reality become so weak** that literally everything we see, wear, use, operate, consume, drive, read, or experience can be **explained away**? That everything within our sight and experience could have come to be *by pure chance, with no purpose*? The foundation of evolution when compared to **reality** or the world that we live in, has **no scientific basis or standard**. When compared to every single other possibility of where things come from, evolution is a most **unlikely answer.**

Nothing, zip, zero, nonexistent, nil, none, all of these words describe how many things are happening on our planet through the process that evolution claims formed the planet and all living things on it! It is equally mind boggling, that some of the same minds could **invent**, and then have others buy into the concept that there was a **big bang**, or an **explosion** that had the power to **organize** a planet, solar system, and possibly the universe, and then, somehow, started all of life! It is like someone took a vacuum and sucked all the light out of view and then filled it back in with black mud! **The concept is pure mythology**! Nowhere at any time throughout all of history or any human experience **has an explosion organized anything** that we know of! *Science is being left in the dust. Science fiction seems to be the Rule*, these **far out guesses** or theories are **not** being promoted by <u>Sound Science</u>. The study of fossils has not produced any **verifiable** facts confirming evolution. There, plain and simply is <u>no scientific **basis** for evolution or the big bang.</u> Neither the theory of evolution or the big bang are ideas based on **reality**! What is Sound Science? It has a deep, strong foundation. It is founded with a **Standard** confirmed by **real evidence**. **Every single thing** with a **verifiable scientific basis** came to be completely in an <u>opposite</u> way of evolution! As we break down evolution, a more clear picture of what it is all about, begins to form.

The Science of Evolution vs the Ideology of Evolution.

The Science of evolution. As I studied evolution I began to understand how one might consider it to have some credibility in scientific research. It is an established fact that within the **layers** of the earth's crust are millions of fossils, and each layer tells a story or gives us a record of that period of time. The earliest or deepest layers show us that the life forms that were in those periods were less complex, such as bacteria or viruses. Even so, these are **not simple** in structure, they are, in fact, extremely complex and **fully developed organisms**. Each outward layer presents new and more complex forms of life, but not more complex in cellular structure, rather more complex in total body structure and also in their influence and domain. No conclusive research so far, has shown us that these creatures in the different layers of the earth are related. In fact, each one **seems** to show us new and different life forms being introduced that have no connection to past species! Those living things are gone now, but each segment or period of time **contributed vital** and **important resources** to our planet. These natural **mineral resources are essential** to all the many things that are happening on our world. Virtually everything we do on this planet depends on them. Step by step, process by process the Creator **built** this planet so it could support all of the living things that are here, with each living thing having all of its needs fulfilled. As all of the evidence has come in, it has quickly shed much doubt on the premise of evolution. But without coming back to the **rock solid** foundation of organized design, which goes against their ideology, core evolutionists <u>had nowhere</u> to go so they simply kept adding more theories, and new angles all built on the same old foundation of sand!

To all Scientists in Biology and related fields: Science research is very important, and the work that scientists do is **vital** to the multiple fields of medicine and those related sciences. This very important work is all done by **intelligent design**! Do you suppose this work would get done **purely by chance** *without you to do it? There is a **definite connection** between the way you do your research and work, and **how <u>all</u>** organized things on our planet came to be. On the other hand the <u>basis</u> or **<u>foundation</u>** of evolution is in direct opposition to everything you are doing,*

The ideology of evolution. There have always been people who simply don't believe in the Creator. They only believe what they can see with their own eyes. They have always sought another explanation of our planet, not accepting the belief that it was created by a Creator. These people have every right to believe whatever they want to believe, this is not the issue. The issue at hand is that evolution has no **scientific basis** whatsoever, and is being taught in our schools **as science** when in fact it is **atheist religion**! In an attempt to legitimize this atheist ideology

65

or religion, they have introduced many theories and built a labyrinth of concepts and ideas while searching the world over looking for evidence of these theories. This **mountain of theories** or guesses, each one **not quite verifiable**, is always followed with more theories or guesses and is a never ending flow, never leading to any solid verifiable conclusion. Yes - *you guessed it* - and is always followed by more theories. This is not being done by the average person who may consider him or her self an atheist. It is being done by a much **more powerful** group of people who are seeking to control the ideology of the world. Why else would we continue down this unverifiable road of dead ends? Evolution then, this atheist explanation of how we came to be, is only a mask hiding what the evolutionists' real goal is - to **take away God** and change the ideology of the world! This is the main reason that evolution came forth to begin with, to be used **as a tool** to remove God from our society very subtly.

Those wanting to change and control the world ideology are only a **portion** of the many scientists, who study in the different fields of biology and earth sciences, but their influence has been huge in setting forth the science agenda in our schools and universities. Unfortunately, all of this research is narrowly set to **prove** that evolution is true at the **expense** of **real science.** Evolution draws away resources and leads us to a dead end!

Most of the **real world** and what is happening here has been **set aside**! Evolution science is in a **total disconnect** with <u>everything</u> that is happening around us!

Virtually **all of the evidence** that is around us, teaches us that **all** things happen, when **intelligent beings** *do something*. That's what we do on this planet! This includes **your own work!** <u>You</u> are an <u>intelligent being</u> working to create the things this world needs! Everything that is happening in your job is happening because **you** are doing it. **Nothing** in the laboratories of the world is happening without **someone to do it!** How can any person conclude that this beautiful and complex world came to be <u>without **someone**</u> to make it? This world is not an anomaly, we are an integral part of the greater universe, we are like one piece of a beautiful, billion piece puzzle. With our better perspective we should be able to see the bigger picture. It's interesting that those who point their finger against God and religion the most, are the ones trying the hardest to promote their own atheist religion.

Chapter nineteen

The Devastating Results of the <u>Ideology</u> of Evolution

Those who promote the ideology portion of evolution and its <u>atheist concepts</u> don't want you and I to know what the results of this agenda will be, a nation **stripped** of our declaration of independence and, as **God is removed, <u>the principles that it stands for</u>**!

We must see that this debate is not one of science vs. religion, it is one of <u>atheist ideology</u> vs. <u>God the Creator of the world</u>. Once again this is where **Real Scientists** need to **Stop and Look** at the consequences or **results** of the theory of evolution. Many have been convinced that the science theories have some credibility, but has the scientific community **focused** on the **very negative effects** to our society that evolution brings with it? Evolution is a very subtle **<u>tool</u>** that is being used to remove God and morality. Many people have simply **lost focus** of what's really happening in our nation. The **policies** promoted by this atheist ideology, are now becoming ever prevalent in our schools and public places, and are attempting to remove the **Creator** out of those places. *Underneath* it all, **that's what evolution does**.

I **am not** saying everyone working in the field. of biology is attempting to remove God and morality from our nation. **On the contrary, most are** very good people trying to do good things for society and the fields of science where they study. Many people are unknowingly being used as **pawns** by this **core** group of atheists to promote their **ideology**. It's that core group we talked about at the beginning of the book who have most of the influence and also set the agenda for many of the science studies that promote evolution.

If we allow God to be removed from America, **black mud** will replace the light that beams from our borders. **God is our Foundation for doing good and being good! America is a great nation because we are Good! If we stop being good we will stop being great.**

History has given us some very good examples of what I'm talking about. When Adolf Hitler came to power he piece by piece *destroyed faith in God* and Religion, then replaced it with the power of the state and the people became *slaves.* Vladimir Lenin in the Soviet Union wiped out religion, while at the same time he **murdered** 27 million of his own people. The rest of the people became *slaves of the government.* China, lead by Mao Zedong, murdered millions of its own citizens, **religion was restricted**, and the people became *slaves of their Godless*

69

government masters! Cuba, North Korea, Laos, and Vietnam. All of these nations are *slaves and servants of their Godless masters.*

We here in the United States of America have been, and are, the main **road block** of those who desire to **destroy freedom on this planet**. So we have **become** the main **target** in the fight to take freedom away from its people. Atheist ideology has always been a **part** of taking freedom from each of the nations above! Evolution is a **subtle persuasion** and the **foundation** of the attempt to **remove God from the world**! They will **never openly admit** their goals to remove the creator and morality from our nation, but the results of their actions are **plain to see** for *those who are **willing** to open their eyes.* The regimes named above killed over **one hundred million** of their own people to establish **their power. Before they could establish their power, they had to remove Morality and God** from their nations.

God is the giver of our unalienable rights, **Not** the government or men! That is why we have **remained a free people** since our nation was founded. As soon as God is removed, and men become the giver of our rights, we as a people will **loose most** of our freedom. **Plain** and **simple – the mission of evolution** is to **remove God! That's exactly what it is doing!** Evolutionists want to separate science from the purpose of life. It simply cannot be done, real science is full of purpose!

Chapter twenty

A Very Short Story

Jake and Lorraine Pumpco had met early in high school and after dating for four years it was just natural to get married and start life together. Jake got a job at an electrical motor rewinding company and learned the business quickly. He could name any brand of motor and rewind it almost in his sleep. So, when five years later the owner decided to retire, he offered the business to Jake. Jake had no problem getting the financing and decided he may as well build a new building on a lot down by the railway line which would enhance his business. That was when Jake ran into Smooth Harry, a small time mob boss. You see Harry wanted that lot, too, and was in the process of getting his money together. When Harry got word that Jake was buying the lot he sent his goons down to lean on Jake, but Jake wouldn't hear of backing down.

Word got around to Hank Manning, a national newsman that Harry had threatened Jake and promised he would be sorry for messing around with him. Hank took out Harry's long file and made note of the threat. He wasn't surprised to see trouble from this small time mobster. Jake went ahead and bought the lot and built a nice building to run his business out of. Several months went by and everything seemed to calm down. Jake hired two fellows who were passing through to help him get going. Jimmy Jones and Tiny Samson had done okay at first, but as time went on Jimmy and Tiny always showed up for pay day but did not put much effort into doing the work.

Finally, Jake had had enough. The two men were not getting their work done, so the time had come to let them go and get some better help. When they were told they were being terminated, Jimmy started yelling and threatening Jake, Tiny simply turned and walked out. The commotion could be heard two blocks away.

Now, to understand this story you must learn about the neighborhood. Across the tracks was an old apartment complex where a number of residents lived. Mrs. Caboss was elderly and loved to watch TV and watch the trains as they rolled on by. In July her windows were usually open.

Mr. Letts was only in his forties but he had been injured and disabled. His hobby was UFO's. He believed aliens were going to take over the planet.

Then there was Eddy, the town drunk. You could see him wander the streets most every day.

Later that night, Jake was in his office finishing some paperwork and was about to go home. Mrs. Caboss heard a loud pop which she took for gun fire. She looked at her clock and made note of the time, 10:30 pm. This is what she would report to the police.

When Mr. Letts heard the pop, he looked out his window and saw a huge crane working a block away lifting a large beam of steel. The crane had many lights on it, and having been awakened from a sound sleep, he thought he was seeing a space ship landing there and ran into the other room to call his UFO friends.

Eddy the drunk had just settled down on Jake Pumpco's dock for a little nap.

The next morning the police received a report that Jake Pumpco had been found murdered. Mrs. Caboss was interviewed and reported hearing gunfire at exactly 10:30 pm.

When Mr. Letts had heard the news, he called his friends who in turn called their friends, etc, etc, and told them aliens had come down and murdered Jake Pumpco and to prepare for an attack.

Hank Manning from the national news wire jumped the gun and sent out a statement declaring he knew that it was too good to be true that Smooth Harry would let Jake get away

73

with taking his real estate lot so he had him murdered. *This media statement went out to the whole country.*

The police went in that morning and began the investigation into what had happened. Jake was found in a pool of blood with a little round hole going right through his heart. The electric clock had stopped at 10:30 pm because the power had gone out. The police found a rubber handled ax on the dock which had been used to cut the electric line causing a loud pop, which would account for Mrs. Caboss hearing what sounded like gunfire. Jake Pumpco had fallen back on his left hand, breaking his watch at 10:43.

Then the police were told Eddy was downtown telling everyone he had seen the murder. Most people didn't pay much attention to him because he was usually drunk, but the night before he had run out of money so he had only one beer and then decided to have a nap on the dock. When the police interviewed him he told them a loud pop woke him up. Jake had come out to the dock to see why the power had gone out and looked around for several minutes. Then out of the dark came Tiny Samson and stabbed him right in the heart.

The police went to the motel where Tiny and Jimmy had been staying, only to find them packed up and ready to take off. However, off to the side of the road in the weeds, they found a sharp, round, steel rod with blood on it. It was soon identified as the murder weapon. A news report was given to the local newspaper which published an article about the murder with all the evidence!

This book is about **evidence,** what is True and what is not true. In this story there were several sources of information.

First, Mrs. Caboss heard what she assumed was gunfire at 10:30 pm.   It was really the sound of the electric wire being chopped by the ax.

Second, Mr Letts told all his friends that aliens were here murdering people. This was absolutely false, no evidence of aliens was found.

Third, National newsman Hank Manning sent out a premature report claiming Jake was murdered by a small time mobster. Thousands of people all across the country had access to that wire. All who read the report had received bad information. No evidence of a mob hit was found.

Fourth: Eddy the drunk witnessed the murder and gave his testimony to the police. The police then captured the murderer and released a statement to the local newspaper stating the evidence.

A few people now had the real evidence and truth about what had happened.

Most of the people who heard about the murder were given **misinformation!**

Information versus Misinformation! This is all about the Truth. What is the Truth?

The Truth is **Knowledge** about what **really happened, and is really happening** with verifiable evidence to back it up.

The issue of what Evolution **really** is, where it came from, and what is the goal of that **core group** of evolutionists, is important to understand. The **ideology** portion of evolution is ice cold. *It steals Human feeling and purpose*, claiming that emotion and feelings are mere biological actions. The subtle undercurrent of the atheist ideology is designed to **destroy any hope** that is given by the Creator, that life continues on and that there is **purpose** to our existence. It **destroys Morality**, (how we react to right and wrong) which, if taken away, would send this world into the *darkest of ages*! Evidence of this is already well documented in history from the above mentioned **godless** regimes.

Evolution is a blast of misinformation, it is being used as a **tool** to steal a belief in the Creator, and morality **as a standard**.

Billions of people believe in the creator and some form of life after death, only a few people firmly believe in the atheist ideology.

Evolution is being **sold** to the world as Science. It simply

76

is **not** science! There is no verifiable evidence of it. No conclusions of evolution **are verifiable**, and **in particular** the **foundation** of evolution has no **scientific basis**!

The Theory of Balance and Harmony, helps to bring back, the "**ability to Reason**" to our Biology Science studies.

The Theory of Balance and Harmony is a theory because it takes us into a place that **we can't** absolutely verify. We can't see the Creator! But this theory presents us with ideas that **are in perfect agreement or harmony** with **everything on this planet**, and **everything that is happening here.** The Theory of Balance and Harmony helps establish, and returns us to, the concept that we have a great purpose in life!

The atheist ideology of evolution presents our nation and the world with a *subtle under current* of destruction.

We must remember where our freedom comes from! Our Nation is built upon the basis of **Unalienable Rights given by God**. These rights are what make our nation Free. **As we remove God**, those unalienable rights will **disappear**! That leaves only **men** to give us our rights, and men as rulers have proven **throughout history,** not to give rights, but to *take away rights*, and attempt to control the people with force!

Once again, evolution **is one of the tools** being used to attempt to remove God, from our public policies and our **society**!

77

Most of the ugliest and darkest things that have happened on this planet have come, at the **hands of men** who have **rejected the Creator** and the unalienable rights that come from Him, to get power, control, and money.

The Belief in a **Benevolent Creator** is the basis for a free society. One that **conforms to moral standards** where laws are made that protect our unalienable rights.

Take a look at this Beautiful World and all that the Creator has given us. It was all made for **our** Experience and Learning! As people of the world we should give this great Creator, Honor Respect and Gratitude!

Chapter twenty one

This chapter is reserved for those who are interested in getting in touch with the Creator.

Warning! It's about Religion

The Creator seems to have been much more focused in his communications with man on **Why** he Created this planet and all things on it, rather than **how** it was done! He commands mankind to seek for Faith in him.

Through modern technology we have learned much about wireless communications. *But we have only just begun to learn, there is much more to know*! The sixth sense or spiritual nature of man is much like a **very high tech receiver.** As the creator organized our spirits He built into them the ability through our feelings and minds to receive from him the help we would need while in this Mortal University! But this is where the **how** things are done of Science, turns into the **why** things were made of **Religion.** Each and every thing around us was made for a **reason! Why** would we believe that there is **purpose in everything around us,** but that **we** came into existence, randomly by pure chance with **no** purpose?

Pure Science teaches us **How** things are made, **Pure** Religion teaches us **Why!**

Each person controls their own spiritual receiver!
Anyone and everyone can tune into or away from the constant broadcast of light and understanding coming from the Heavens. This is why there is such a dramatic difference in the results of each person. Some are **very** fine tuned, others are **completely tuned out,** the rest of us are somewhere in the middle! Those who are tuned out of this broadcast make the cry, where is this unknown God? **Well, it's all in the tuning!** Many people get hung up on these things because they cannot **see** with their eyes, or **hear** with their ears, or **feel** with their hands. I would challenge you to learn about your spiritual nature so you can realize just how powerful it is. **Your Feelings are your most powerful sense!** (When mind and heart are in Balance and Harmony). Now as for the answer for those who want to make contact with the Creator, **everyone** who <u>honestly</u> seeks the Creator on the Creator's terms can find him. <u>On his terms</u> is the KEY. This also is pure science, just as a radio station broadcasts on a frequency out to any radio that is **<u>tuned</u>** to that station. The Creator, who made our spirits, built in this automatic spiritual receiver, and is constantly broadcasting to all who are <u>tuned in</u> (on His terms) to his broadcast.

As each of us focus on, and realize how much we use our sixth or (spiritual) sense, we will then see that tuning into that vast pool of knowledge coming from the heavens will help us gain a greater understanding of **our purpose for being here and our great potential**. Why have the **minds of men been plunged into darkness concerning evolution?** Because **Faith is getting very weak.** Not only **Faith in the Creator,** but faith in what each person can see with **their own eyes** and in what they are **actually experiencing**. Consequently evolution entered the scene!

The Creator is **Very Interested** in what is going on here in this great mortal experience. <u>**He**</u> is the one who **made us** *for a great purpose.* So does each and every thing around us have a purpose. What is that purpose? To enable each and every creation to reach its **highest and best Potential**!

For example, when a baker makes a cake he wants it to be eaten and enjoyed. Virtually everything that people make has a **reason** for being made. When we make something we want it to be the **best that it can be** or at least to fill some good function.

Mankind is the Creator's very best Creation, in fact we are made in his image and are his spirit children! He desires that each person reach their **full potential**. This experience on this planet is **custom designed** for that purpose!

It will take Courage, Faith and never giving up in the face

of what sometimes may seem very difficult or unbeatable odds. *Unfortunately **not everyone** will **choose** to accomplish this*!

All other things including this planet were **made to aid us** in reaching **our** potential, while at the same time reaching their own highest and best. There are a few keys that we must learn, to get through this experience.

First, the Creator **does not use force** to make us do anything, <u>we get to choose</u>!

Second, as we choose and then act, we receive the **consequences** of our choices - <u>good or bad</u>!

Third, but not least, God the Creator is **always** there to help us as we seek him, **(tune him in)** and then try to **follow** him. He is not dead nor does He sleep, He is actively involved with us.

Fourth, but First, and not least, but most, that we learn to Love God, and Love our neighbors as ourselves. (How we treat others). (Loving God may seem difficult, but as we learn the **<u>Truth</u>** of all he has done for us, our Love and Respect will come much more easily). **Indeed He deserves Great Honor and Respect.** These are **just the basics**, but as we learn more about **Pure Religion, (pure religion** is unbiased and based on **Truth** rather than the traditions and **ideas of men**), we will grow to understand more of why we are here! The Creator fills the universe with **<u>positive</u> <u>power</u>**, it's up to us to **choose** to connect with that power. Now we can see why atheists hate the concept of "Creation with Purpose". They have long ago **tuned out** the constant broadcast coming from the heavens and have chosen their own path. Those people who use their agency to rebel against the Creator, create a **negative energy,** which accounts for the selfish, mean, hurtful, and violent things that happen on our planet! The battle between the **Positive of the Creator** and the **negative of** all those who **rebel**, has been going on since the beginning of our creation, when the Creator gave us the **agency . to choose**! (or ability to choose). As we choose, we receive the *consequences of our choices*. When people in any *family, nation, or community willingly* **choose to disregard** the Creator or His natural laws, it is **only natural** that the negative effects or *consequences* will take their course against that entity. These are the natural laws that govern this earth and all of space. Why does Positive **always** win **in the end**? Because each and every **negative and immoral behavior** eventually works to **destroy itself,** as dishonest, or cruel people move in a self-destructive course. **Morality matters very much;** our choice, is to follow God's guidance, **or not**. The Creator's laws are Eternal in nature, we don't get to pick and choose which ones we like or don't like, will follow or not follow, without receiving the consequences of our choices.

His laws are **designed** to **lead** us to happiness! We cannot disregard these **laws** of the Creator, which are the same as the Natural Laws that govern the universe, and then find happiness!

The positive of God will **always** maintain and strengthen Freedom and happiness. The negative, of opposing God will **always** addict, ensnare, enslave, fool us, bind us down and **remove** our Freedom and happiness.

It's **not one** big choice but all of the **many little ones** we make each day. The Purpose of our loving **God** is **not** to fulfill *all* of the desires that we have, nor to **agree with us** when we **choose** to veer off course, **because He won't;** but to help **us** **choose** Positive over negative, good over bad, and self-control over selfishness or being self-centered. **Our choices** will **take us** to where we will be throughout Eternity. Through making good choices we **avoid negative consequences.** Remember: the Saviors Love and help are **always** available, He is the Redeemer of mankind, and every single person on this planet has the ability to **tune in** to this great power coming from the Heavens. The Creator is no respecter of persons. He does not care about race, creed, or situation, only the **purity** of ones **heart** and **desires,** and **whom** we **choose** to follow! This is what makes up the **Love of God and Pure Religion!** Just **remember**, we are part of a Great and Beautiful Plan and **you** have Great Purpose!

www.ingramcontent.com/pod-product-compliance
Lightning Source LLC
Chambersburg PA
CBHW071246170526
45165CB00003B/1254